Measuring Tools

by Mickey Daronco and Lori Presti

Table of Contents

Which Tools Do We Use
to Measure Length? 4
Which Tools Do We Use to Measure Weight? ... 8
Which Tools Do We Use
to Measure Volume? 10
Which Tools Do We Use
to Measure Temperature? 12
Which Tools Do We Use to Measure Time? 14
Glossary and Index 16

We use tools to measure things. Here are some of the tools we use when we measure. Different kinds of tools measure things in different ways.

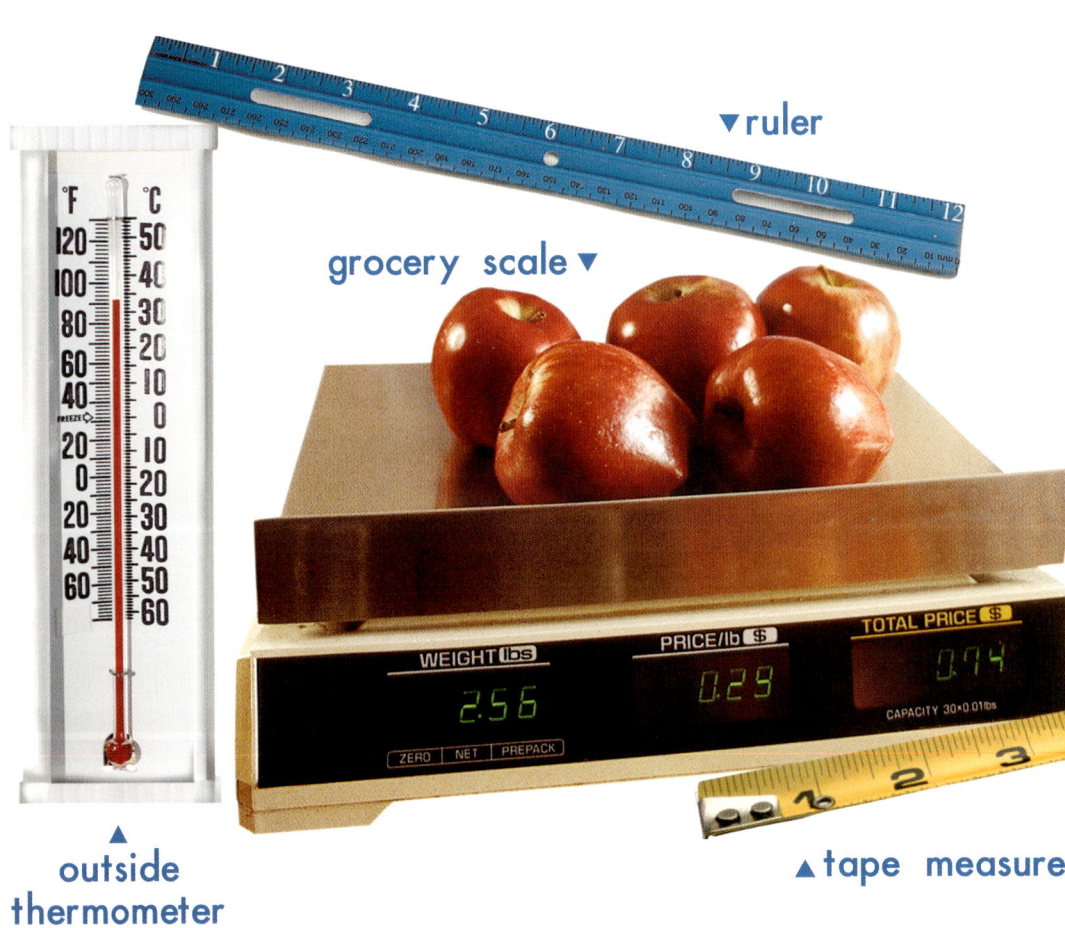

ruler ▼

grocery scale ▼

▲ outside thermometer

▲ tape measure

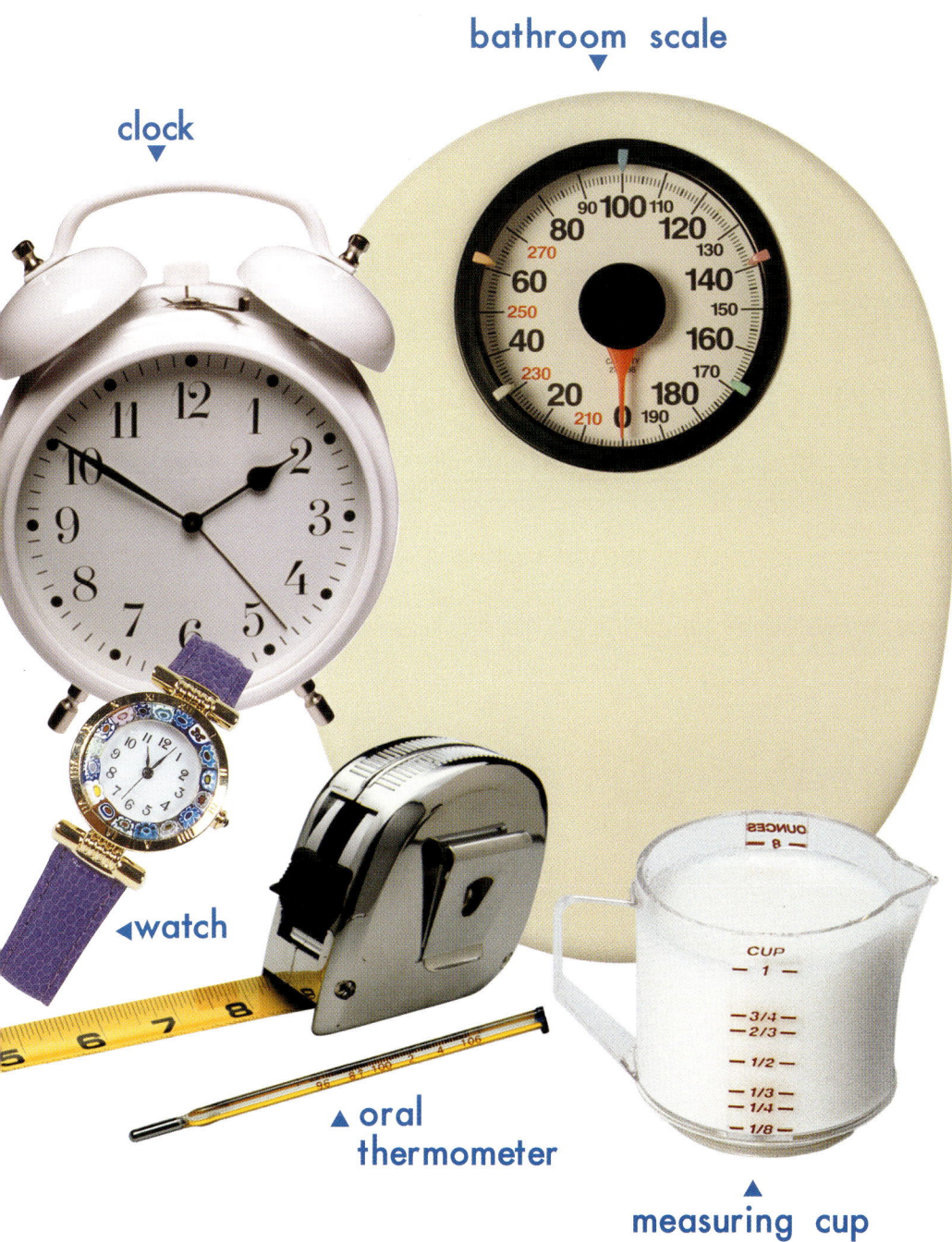

Which Tools Do We Use to Measure Length?

This girl wants to make a card for her mom. She needs to cut the paper in half. Which tool should she use to measure the paper?

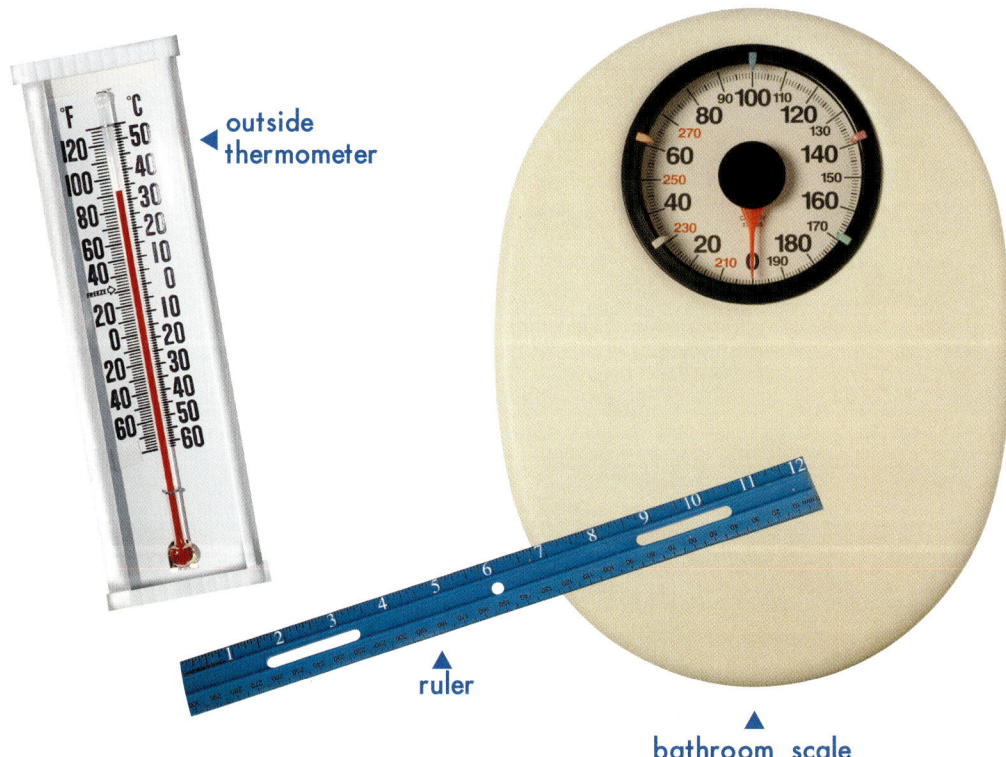

outside thermometer

ruler

bathroom scale

Rulers measure length, or how long something is.

This man is building a house. He wants to buy some lumber. He needs to know how long the lumber is. Which tool should he use to measure the lumber?

clock ▶

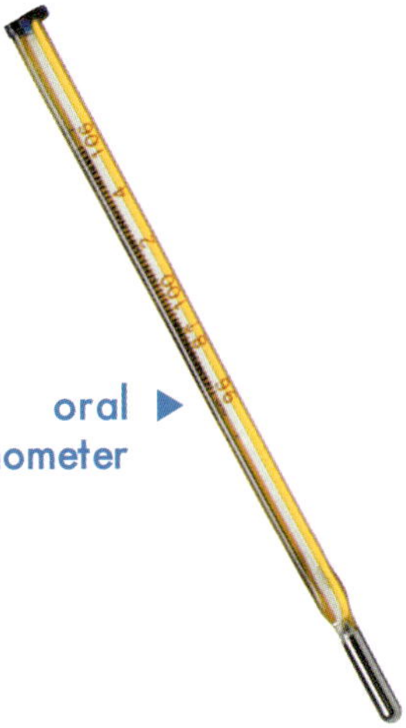

oral ▶
thermometer

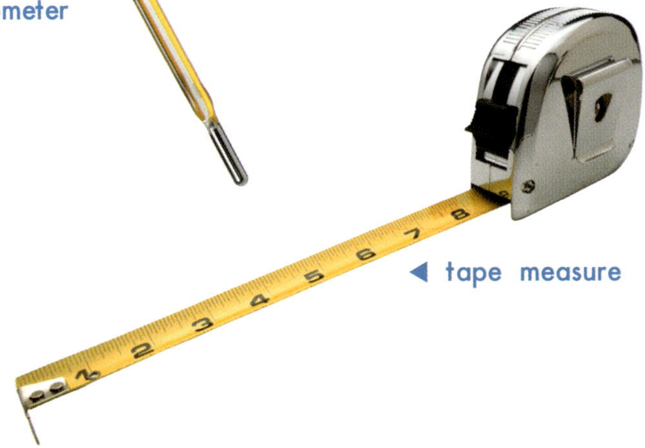

◀ tape measure

When we measure things that are long, it is easier to use a tape measure than a ruler. A tape measure is longer than a ruler.

Which Tools Do We Use to Measure Weight?

▲ grocery scale

This woman wants to buy some cheese. She is going to make a lot of sandwiches. Which tool should the grocery use to measure the cheese?

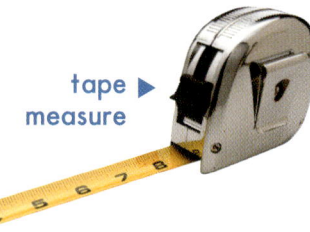
tape ▶ measure

watch ▶

Grocery scales measure weight, or how heavy something is.

This girl thinks her cat has gotten fat. She wants to know if her cat is heavier than it was last year. Which tool should she use to measure her cat?

ruler

watch

bathroom scale

A bathroom scale measures weight, too.

Which Tools Do We Use to Measure Volume?

This baker is making a cake. He has put the flour, sugar, and eggs into the bowl. Now he needs to add the milk. Which tool should he use to measure the milk?

◄ clock

measuring cup ▼

▲ grocery scale

A measuring cup measures volume,
or how much space something takes up.

Which Tools Do We Use to Measure Temperature?

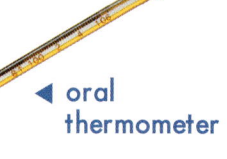
◀ oral thermometer

This girl is getting ready to walk to school. She wants to know if she should wear a coat. She needs to find out if the temperature outside is hot or cold. Which tool should she use to measure the temperature?

▲ watch

▲ outside thermometer

An outside thermometer measures the temperature outside.

This boy does not feel well. His mom wants to know if he has a fever. Which tool should she use to measure his temperature?

outdoor thermometer ▶

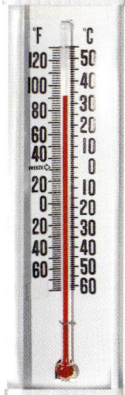

oral thermometer ▶

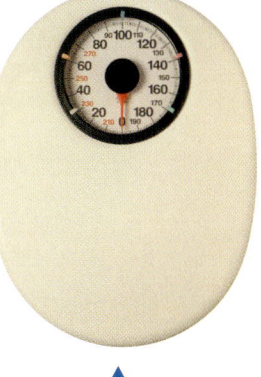

▲
bathroom scale

An oral thermometer measures the temperature of the body.

Which Tools Do We Use to Measure Time?

This girl is having a friend come to her house after school. She can't wait to go home! She wants to know how much more time she will be in school. Which tools could she use to measure the time?

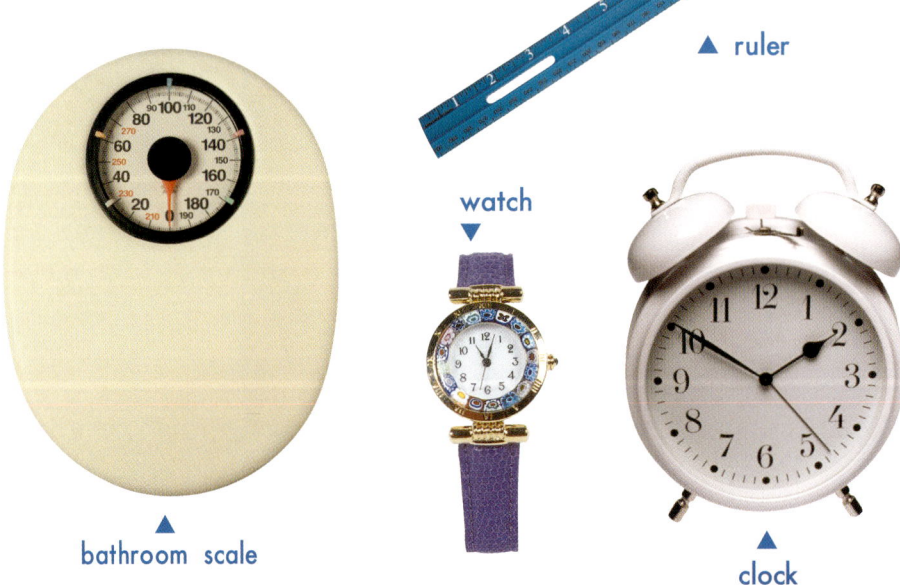

▲ ruler

▼ watch

▲ bathroom scale

▲ clock

Clocks and watches measure time in hours, minutes, and seconds.

Glossary

clock (KLAHK): a tool that measures time
length (LENGKTH): how long something is
lumber (LUM-ber): wood
measure (MEH-zher): to find the size, weight, or amount
measuring cup (MEH-zhuh-ring KUP): a tool that measures volume
ruler (ROO-ler): a tool that measures length
scale (SKALE): a tool that measures weight
tape measure (TAPE MEH-zher): a tool that measures length
temperature (TEM-puh-ruh-cher): how hot or cold something is
thermometer (ther-MAH-meh-ter): a tool that measures heat or cold
tool (TOOL): something that helps people do work
volume (VAHL-yoom): how much space something takes up
watch (WAHCH): a tool that measures time
weight (WATE): how heavy something is

Index

clock 3, 6, 10, 14–15
cup 3, 10–11
length 4–5
lumber 6
measure 2, 4–15
milk 10

ruler 2, 4–5, 7, 9, 14
scale 2–4, 8–10, 13–14
temperature 12–13
thermometer 2–4, 6, 12–13
time 14–15

Glossary

clues: (KLOOZ) hints that help to answer a question.
guess: (GES) to say or think, without knowing for sure.
hippo: (HIH-poh) a short word for hippopotamus.
less than: (LES THAN) smaller than.
more than: (MOR THAN) bigger than.
number line: (NUM-ber LINE) a line with numbers on it, like 1, 2, 3, and so on, to the end.
secret: (SEE-kret) something hidden that not everybody knows.

Index

clue: 8, 14
guess: 2–7, 10, 12–13, 15
hippo: 2–3, 5–6, 11, 13–15
less than: 8, 14
more than: 14
number line: 3
secret: 2, 4, 8, 10–12, 15

She guessed that my secret number was 11. She was right!

Were you?

You should try playing Two Hungry Hippos with your friends. You can have a lot of fun and learn about numbers at the same time.

I told her my number was smaller than 14. She moved the green hippo down to cover 14. Then I gave her a clue.

I told her that my number was 3 less than 14 and 3 more than 8. Now she said she knew what my number was.

She moved the red hippo up to cover fish 4. My friend guessed 14.

What would you guess?

What Was My Secret Number?

I wrote down my secret number. My friend guessed that it was the number 4. I told her my number was bigger than 4.

This was fun. Now it was my turn to think of a secret number.
We moved the hippos back to the ends of the line.

I guessed that her secret number was 10.

I was right!

Were you?

My friend said that I was getting close. She said that she would give me a **clue**.

She said her number was 2 **more than** 8 and 2 **less than** 12. Now I was sure I knew what her secret number was.

Do you know what it is?

Now I knew that her number was between 2 and 12. I guessed the number 7.

What would you guess?

My friend told me I'd have to guess again. She said that her number was smaller than 12.

I moved the green hippo down to cover fish 12.

I moved the red hippo up to cover fish 2. I guessed her number was 12. What would you guess?

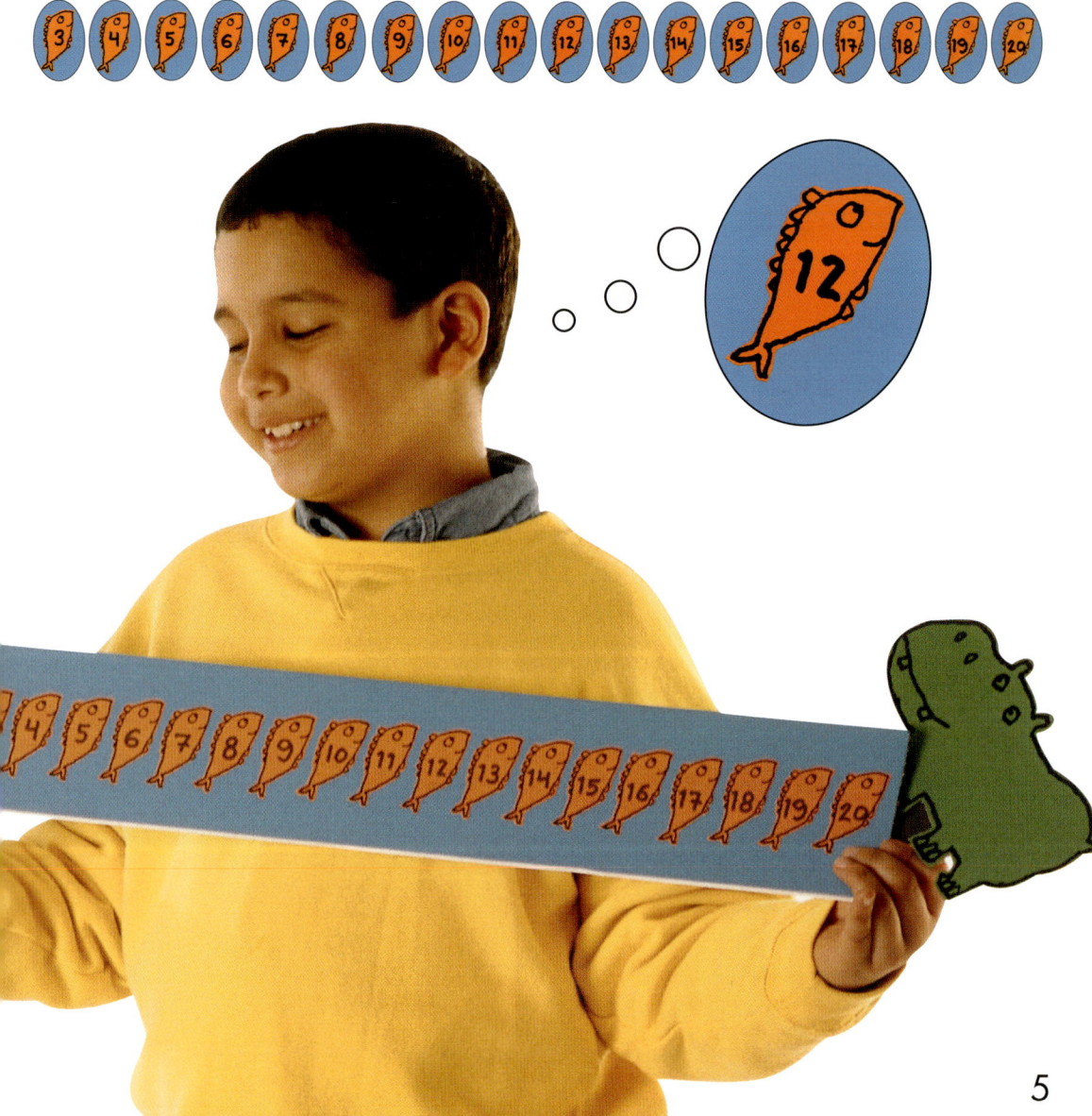

What Was My Friend's Secret Number?

She said she was thinking of a secret number between 1 and 20. She wrote it down. I guessed that it was the number 2.

She said her number was bigger than 2.

She took out a big **number line** with many fish on it. Each fish had a number on it from 1 to 20.

A hippo was at each end of the line.

She told me I could move the hippos to help me guess her number.

What Is the Two Hungry Hippos Game?

My friend wanted to play the Two Hungry **Hippos** game. She said that she would think of a **secret** number. Then I would have to **guess** what her number was.

Two Hungry Hippos Math Game

by Alison Adams

Table of Contents

What Is the Two Hungry Hippos Game?........ 2
What Was My Friend's Secret Number?........ 4
What Was My Secret Number?............... 12
Glossary and Index...................... 16